This book is dedicated to all who find Nature not an adversary to conquer and destroy, but a storehouse of infinite knowledge and experience linking man to all things past and present. They know conserving the natural environment is essential to our future well-being.

CANYONLANDS

THE STORY BEHIND THE SCENERY®

by David W. Johnson

David W. Johnson, an associate professor of biology and environmental science at The College of Santa Fe, New Mexico, received his Ph.D. in ecology from the University of Colorado at Boulder. He served seven years as seasonal ranger and biologist at Canyonlands National Park. David is also the author of *Arches: The Story Behind the Scenery.*

***Canyonlands National Park,** located in southeastern Utah, was established in 1964 to preserve a geologic wonderland of rocks, spires and mesas, and a variety of Indian rock art.*

Front cover: Angel Arch; Inside front cover: The All American Man, up Salt Creek in the Needles District, photos by Jeff Gnass. Page 1: Needles Skyline, photo by Peter Kresan. Pages 2/3: Caterpillar Arch in Lavender Canyon, photo by Becky & Gary Vestal.

Edited by Mary L. Van Camp. Book design by K. C. DenDooven.

Seventh Printing, 2005 • New Version

LC 89-45018. ISBN 0-88714-034-3

Imagine an arid land of colored rock;
a land of cliffs, canyons, arches, spires,
and mesas carved by water and time;

a land of grand vistas and narrow,
rocky crevices; a wild land visited by
people for millennia—Canyonlands.

The Canyonlands Story

ALEXANDER SKYE

Banded spires of Cedar Mesa Sandstone *such as those that line Elephant Canyon are the source of the name of the Needles District. With advances and retreats of a shallow sea during the Permian Period, alternating bands of white and red deposits formed.*

The Colorado Plateau is a rugged land, a land of contrasts—unbroken expanses of rock, towering monoliths, great rivers, scorching canyon bottoms, and icy mountain peaks. At the heart of the Plateau is a mile-high maze of canyons, mesas, and spires, carved by the Green and Colorado rivers—Canyonlands National Park. This land of canyons is at first glance barren and stark. A high-desert land of rock as far as the eye can see. Intimidating and forbidding. But look more closely. Life exists, even thrives, among the rocks. Hanging gardens support columbine, monkeyflower, and fern; cliffs and rubble slopes are home to mule deer and bighorn sheep.

People lived here long ago. Little-known desert ancients hunted the canyon country and recorded their passing on the cliffs. Anasazi Indians farmed the meager bottomlands of the canyons. Cowboys roamed twisted canyons seeking stray, half-wild cattle. Now we are here, visitors to this bold, quiet land. Sit in the shade of a boulder or piñon and let Canyonlands come alive.

Life exists, even thrives among rocks

Paul Bunyans Potty is a pothole arch *in Horseshoe Canyon. Water scouring a depression on the cliff tip cut through the rock and joined the underlying alcove to form a horizontal or pothole arch.*

BECKY & GARY VESTAL

JEFF GNASS

The "Four Faces" *pictograph panel is one of several of this style in the Needles. Faces or masks were painted in red or white pigment; at one site, the rock was smoothed before the face was painted.*

Green and Colorado rivers are the major agents
wo powerful sculptors
join in the center of the park and divide it
into three sections—the Needles,
Island in the Sky, and Maze districts.

Land of Layered and Sculpted Rock

DAVID MUENCH

Caps of erosion-resistant White Rim *Sandstone protect the softer underlying reddish-brown Organ Rock Shale and create a collection of spires in the Island in the Sky District.*

KEITH GUNNAR

High above the Colorado River, a lone *figure stands atop Cedar Mesa Sandstone and surveys millions of years of geologic history.*

Water, gravity, and time have hewn the modern landscape from an ancient rocky foundation. Perhaps surprisingly, in this arid land, water is the major agent of erosion. The importance of water's role in shaping the land should not be underestimated. Rivulets and pools dissolve the cement that holds

GARY LADD

***From the White** Crack campsite on the White Rim, the La Sal Mountains rise to over 12,000 feet on the eastern horizon. Like the Abajo Mountains to the south of Canyonlands, the La Sals formed when deep molten rock moved upward and bowed the overlying rock strata.*

sand grains together. In winter, water freezes and thaws and frees fragments of stone. Bit by bit erosion slowly etches new patterns on the landscape.

At times, the pace of erosion is quick and dramatic. Flash floods tear through canyon bottoms, moving boulders, undercutting cliffs, and carving into the canyon floor. Spring flooding of the Green and Colorado rivers rearranges sandbars and scours ever deeper into the riverbed. Slabs of sandstone peel away from vertical cliffs and crash to the canyon bottom. Let us look at the forces that have shaped Canyonlands.

From the center of the park at Grand View Point or Confluence Overlook, layer-upon-layer of sedimentary rock sit like an intricately carved layer cake. From the ancient Paradox Formation to the graceful domes of Jurassic Navajo Sandstone atop the Island in the Sky, the geologic strata reveal a record of past environments.

Water and time have already carried away a mile-high thickness of rock that once overlaid the land. What we see now is the current stage of continuous erosion. In our lifetimes, little change in the rocky landscape will be evident. Yet, erosion is continuing and little by little Canyonlands is being carried to the sea.

Origins of this landscape lie in the characteristics of the sedimentary rock formed during the Pennsylvanian geologic period 320 million years ago. At that time, towering mountain ranges arose in the Rocky Mountain region. One of the several 15,000-foot-high ranges was the Uncompahgre Uplift, a sierra that cut across western Colorado and through Utah just north of present-day Canyonlands.

An adjacent and equally great subsidence or basin just to the south accompanied this uplift. The Paradox Basin, as it is now called, was intermittently connected to the sea on the west and south. During its 30-million-year existence, this basin alternately became very shallow or even dried, then recharged with fresh loads of sea water. Twenty-seven of these cycles of connection, then isolation from the sea resulted in evaporation of water and accumulations of previously dissolved minerals. Huge deposits of "evaporites" such as potassium chloride and sodi-

PETER L. KRESAN

***The colorfully banded spires of the Needles** District are formed of Cedar Mesa Sandstone, part of the Permian Cutler Formation. Faults and joints in the underlying formations cracked the Cedar Mesa Sandstone into a complex pattern of massive rectangular blocks. When this layer was exposed, erosion widened many of the cracks and created a profusion of spires—some tall and slender, some short and squat. Between the needles of rock, narrow clefts sometimes form an intricate network of interconnected passages.*

um chloride (table salt) built up in the basin. At the same time, debris from erosion of the nearby Uncompahgre Uplift added dark bands of shale to the layers of salts. In the present Canyonlands area, this Paradox Formation is more than 3,000 feet thick.

The salts of the Paradox Formation played a vital role in forming today's scenery. As we shall see later, the plastic or "squishy" nature of salt influenced the shape of the rocky landscape above.

As the Pennsylvanian Period drew to a close, the ancient sea was receding from the canyon country and a mixture of marine and terrestrial sediments was laid down. Little salt was deposited, but sea sediments mingled with debris from the Uncompahgre Uplift and created the Honaker Trail Formation, the only Pennsylvanian rocks visible in the park. Cliffs of this formation are the foundation of the canyon of the Colorado River as it cuts through Canyonlands.

The colorful, fantastically carved spires and massive cliffs that are the essence of Canyonlands developed from layers of rock deposited during the 100 million years following the Pennsylvanian Period—the Permian, Triassic, and Jurassic geologic periods. The transition between the Pennsylvanian and Permian periods was not dramatic; the earliest Permian deposit somewhat resembles the Honaker Trail Formation. An arm of the ancient sea reached into Canyonlands and created a thick layer of limestone—the Lower Cutler beds. Fossils of marine creatures are found here.

Above the limestone lie the red and white bands of the showy Cedar Mesa Sandstone Member of the Cutler Formation. Canyonlands was in a transitional zone between dry land and shallow sea during the Permian Period. Erosion of the Uncompahgre Uplift to the north and east brought reddish deposits, while beaches and shallow sand bars deposited light sediments. With small retreats and advances of the sea, alternating beds of red and white sandstone formed and today are the basis of the multicolored formations in the Needles and Maze districts.

PETER L KRESAN

***200 million years ago winds** piled massive sand dunes that now comprise the Wingate Formation. This formation averages 300 feet thick in the park and forms a nearly impenetrable barrier to travel. Here, cliffs of Wingate Sandstone guard the entrance to Upheaval Dome and form the bulwark of the Island in the Sky mesa.*

The *setting* sun highlights the colors of the canyon *rock*

During one retreat of the sea, layers of silt, mud, and sand were left by freshwater streams on top of the Cedar Mesa Sandstone in the Island in the Sky area. These layers formed the red Organ Rock Tongue component of the Cutler Formation, visible from Grand View Point. A final invasion by the sea during the Permian Period left a hard, 200-foot-thick layer of white sandstone called the White Rim. This most recent member of the Cutler Formation is aptly named because it forms a broad bench on the Island in the Sky between the rivers and the cliff tops.

In the course of the next 40 million years, the Triassic Period, the fingers of the western sea occasionally lapped over the land and left thin layers of marine sandstone and limestone. During much of this period, however, Canyonlands was a broad river delta with streams that deposited the shales, siltstones, mudstones, and sandstones of the reddish-brown Moenkopi Formation. Above the Moenkopi lies the Chinle Formation, a sandy and silty shale left behind by freshwater streams and lakes. Uranium-bearing beds are often part of the Chinle.

Toward the end of the Triassic Period, Canyonlands and much of Utah was desert, the classic desert we envision when we think of the Sahara or the Arabian deserts. Thousands of square miles of sand dunes, blowing and shifting in the winds, covered much of the Colorado Plateau. These arid sand hills gave rise to the spectacular cliff-forming sandstone that encloses much of the park—Wingate Sandstone.

Atop the majestic Wingate escarpment, sit the sandstones and shales of the Kayenta Formation. Near the end of the Triassic, the waterless land of blowing sand gave way to freshwater streams and lakes originating to the west. The sturdy, hard nature of the Kayenta Formation protects the underlying Wingate Sandstone from erosion, resulting in the massive vertical cliffs characteristic of the Wingate.

The youngest rock remaining in Canyonlands, and visible only on the Island in the Sky, is Navajo Sandstone. Like the Wingate, Navajo Sandstone originated in expansive sand dunes. As in Triassic times, desert also characterized the Jurassic Period; blowing sand covered most of Utah. This return to sand-dunes-dominated aridity resulted at least partly from the building of a mountain range hundreds of miles to the west which intercepted most of the rainfall that had previously fallen on the Plateau.

While the geologic layer cake ends here in

Water and gravity are the two powerful erosional *forces that relentlessly carve the land. Here, poised on the brink of collapse, a block of White Rim Sandstone sits atop an ever-shrinking pedestal of Organ Rock Shale. It could topple tomorrow—or stand for centuries.*

GARY LADD

The setting sun highlights the colors of *the canyon country rock. The White Rim reflects gold and the cliff-forming Wingate Sandstone of Junction Butte turns to a warm reddish-brown. As erosion removes the Organ Rock Shale, chunks of the overlying White Rim Sandstone break away and dot the rubble slope*

GARY LADD

***Storm clouds build over** the Doll House in the Maze District Paradoxically, in this arid land water is the chief agent of erosion. Rain and melting snow dissolve the cement that binds grains of sand into rock; cycles of freezing and thawing pry slabs of stone from cliff faces.*

DIANE ALLEN

Canyonlands, another 140 million years of geologic history occurred after our Canyonlands record disappears. At least a vertical mile of rock strata deposited in subsequent periods is now missing because the mighty Colorado and Green rivers and the inexorable forces of erosion stripped away these layers. In the regions surrounding Canyonlands, many of these younger formations remain. Arches and Bryce Canyon national parks preserve more recent formations such as Dakota Sandstone and Mancos Shale. Grand Canyon National Park and Dinosaur National Monument exhibit many older deposits; formations not yet exposed in Canyonlands.

The Green and Colorado rivers are the major agents of erosion. These two powerful sculptors join in the center of the park and divide it into three geographic and administrative sections—the Needles, Island in the Sky, and Maze districts. Their similar geologic histories unite the districts, but each has its own geologic "personality."

The Island in the Sky

The Island in the Sky is a land of mesas and sweeping vistas. In two giant, thousand-foot steps one ascends from the bottom of the river canyons at an elevation of 4,000 feet to the bench of White Rim Sandstone, then up to the top of the Island in the Sky at over 6,000 feet. A paved road enters the park at its northern boundary and runs atop the Kayenta Formation and Navajo Sandstone to Grand View Point. Here, stretched out before you is a vast panorama of over 10,000 square miles of rock, canyon, and mountain. Off to the east, rise the La Sal Mountains. Far to the west, above the Orange Cliffs of Wingate Sandstone, Thousand Lake Mountain and the Henry Mountains form the western horizon. The Book Cliffs shimmer just out of sight 60 miles to the north. Though the view south is partly blocked by nearby towering Junction Butte, to the southeast the Abajo Mountains and Elk Ridge provide a forested horizon.

A dozen or so miles to the southeast, the red-and-white banded spires of Cedar Mesa Sandstone punctuate the skyline of the Needles District. The same distance to the southwest leads the eye to the tortuous canyons of the Maze District. Both areas are nearby as the raven flies, but for automobile travelers it is a three-hour drive to the Needles and a

TOM TILL

The jagged core of *Upheaval Dome reveals a jumble of White Rim Sandstone ringed by slopes of Moenkopi and Chinle formations which rise to Wingate and Kayenta cliffs. The origin of this crater-like structure is uncertain.*

marathon all-day car and jeep drive over rough roads to the Maze. Distances here are deceiving. Canyons, rivers, and cliffs inhibit foot travel and few roads penetrate into the rocky wilderness.

Just below Grand View Point the elements have gouged the White Rim Sandstone bench, exposing the underlying Organ Rock Tongue. Water, frost, and gravity have carved towers and spires from the soft shale, and 300-foot-tall monoliths stand like sentinels in the basin.

Several unique geologic features add to the spectacular scenery of the Island in the Sky. Closer to the Green River lies what looks like an enormous blemish on the surface of the land. Upheaval Dome resembles a mile-wide volcanic crater with slopes formed of tilted layers of Moenkopi, Chinle, Wingate, Kayenta, and Navajo formations. At the center of the pit, White Rim Sandstone forms several little pointed peaks.

Although Upheaval Dome is not an ancient volcano, there is some disagreement over how it formed. Recall the massive salt deposits of the Paradox Formation. When thick layers of salt are squeezed, the material becomes plastic or malleable and tends to flow. Here, thousands of feet of rock overlying the salts pressed down on the deposits, and, according to early geologists, salt oozed upward at a weak place and formed a dome of salt that has since collapsed and eroded into a crater.

More recent study has shown that the Dome is possibly an eroded astrobleme—a meteorite crater. Perhaps, 65 million years ago, a meteorite one-third of a mile in diameter crashed and what we see today is the remnant of this impact. The rocks now exposed were buried beneath several thousand feet of overlying formations at the time. Erosion of those layers has revealed what remains of the collision.

The *terrestrial* **canyons** were **models** for **geologic** processes ***on Mars***

THE NEEDLES

On the Island in the Sky you survey Canyonlands from above. The scenery falls away from your feet and sweeps to distant horizons. In the Needles and Maze districts, by contrast, the landscape surrounds you. You walk among the spires and canyons that you saw from above on the Island. The scale is easier to appreciate. You can stand beside a cliff or tower and sense the texture of the rock, feel the coolness remaining in early morning or the barely tolerable heat after the summer sun has baked the rock. There is an intimacy between visitor and landscape that is possible only when we can see the land up close.

The combination of ancient geologic events and ongoing erosion by water, frost, and gravity carved the fantastic cliffs, spires, and arches in the Needles and the Maze. Removal of thousands of feet of rock that once overlay the colorful Cedar Mesa Sandstone gave erosion the opportunity to sculpt. But events

JOSEF MUENCH

"Wherever we look there is but a wilderness of rocks; deep gorges, where the rivers are lost below cliffs and towers and pinnacles; and ten thousand strangely carved forms in every direction; and beyond them, mountains blending with the clouds." (John Wesley Powell in "Exploration of the Colorado River of the West and Its Tributaries.")

***Dissolution and flow of deep** deposits of salt—part of the Paradox Formation—stretched and deformed the overlying rocks and formed long, parallel faults. Enormous rectangular blocks of rock enclosed by these fractures sank and formed the shallow, straight-sided valleys called grabens. This aerial view shows more than a half-dozen in the Needles District. Superimposed on the ribs between the grabens is the pattern of right-angle cracks that dissects much of the land in the district.*

far below in the Paradox Formation had dramatic impacts on the shape of the overlying rock.

Sixty million years ago, the land just to the south of Canyonlands arose. The northern flanks of this uplift—the Monument Uplift—tilted down toward the junction of the Green and Colorado rivers. Deep underground, water drained down the slope of the uplift and, along the way, dissolved some of the salts of the Paradox Formation. With some of the underlying support gone, overlying formations sagged and cracked; faults and fractures in the rocks laid a basic pattern upon which erosion works.

In addition, the enormous weight of the overlying rock pressed down upon the salt and "squeezed" it downslope toward the Colorado River. The flow of salt stressed and stretched the overlying rocks and resulted in a series of parallel faults. Looking down from an airplane, the pattern of cracks and fractures in the Needles fault zone is remarkably clear. Faults radiate from Cataract Canyon in a series of arcs. Long rectangular blocks of rock sank and formed straight-sided, flat-bottomed canyons, called grabens. Because these are relatively young structures, their vertical walls have eroded little; while in a few, watercourses have no drainage outlet.

In the 1970s, geologists studied the grabens in Canyonlands because similar features had been photographed on Mars. The terrestrial canyons were models for geologic processes on the red planet.

KEITH GUNNAR

Some of the hiking *trails in Canyonlands are no more than routes marked across the sandstone. Great expanses of slickrock are natural paths between canyons. The blackish coating on some of the rock is desert varnish—mostly clay, iron oxide and manganese oxide deposited by wind and water.*

DAVID W. JOHNSON

At the head of Elephant Canyon, Druid *Arch towers above the canyon bottom and resembles the rock structures at Stonehenge. The only access to the arch is by foot: along trails from Squaw Flat Campground, Elephant Hill, and Chester Park. The arch may have first been viewed in modern times by cowboys David Goudelock and his father in 1915.*

GARY LADD

The massive Cedar Mesa Sandstone near Squaw Flat *mirrors the yellow and gold leaves of the cottonwoods in fall. Deposits of sand dropped by the wind and washed from the cliffs have accumulated in the flats and support scattered piñon and juniper trees, shrubs, and grasses. Just as the color of vegetation changes with the season, the hue of sandstone appears to change by the hour.*

***The Land of Standing Rocks is** at the heart of Canyonlands. It is remote, rugged, and intriguing. Although people have visited this land for thousands of years, its wilderness lets us feel as though we are the first to explore its canyons and mesas. It is not easy territory to explore, but the rewards are magnificent.*

The Maze

The Maze District of Canyonlands lies west of the Colorado and Green rivers. Few visitors to the park get more than a tantalizing glimpse of this remote region from Grand View Point, Green River Overlook, or Confluence Overlook. Place- names such as Land of Standing Rocks, Doll House, The Fins, and The Maze vividly describe the geologic scenery.

MIKE HILL

The Maze is a complex, tortuous system of canyons that is nearly inaccessible. Only a few trails lead into the labyrinth carved into Cedar Mesa Sandstone. As in the Needles District, the Cutler Formation, especially Cedar Mesa Sandstone and Organ Rock Shale, is the foundation of the geologic scenery in the Maze. Joints and fractures have cracked the rock. Water and gravity have whittled fanciful shapes: The Wall, Lizard Rock, Chimney Rock, the Chocolate Drops, and Beehive Arch are massive stone portrayals of their namesakes.

The Needles and Maze are lands of innumerable canyons and spires. Countless tributaries of the Colorado and Green rivers arise in the higher elevations of the park and as they lead downstream they join forces and form ever larger canyons. Except for a scattering of springs and seeps, most of these canyons and side canyons are usually dry. Water flows down the sandy and rocky courses only after rain and snowstorms.

Salt Creek, for example, is usually dry and sandy at its lower reaches, but a cloudburst over its headwaters can unleash a six-foot-deep deluge. These flash floods sweep away boulders and trees, and rearrange the canyon bottom. A bit of the land is carried into the Colorado River. Floods are common but short-lived events and may alter the l and only a little. But when the effects of floods combine with frost, wind, gravity, and enormous amounts of time, great canyons are created.

Our human perspective of time makes it difficult to appreciate the power of erosion in a dry climate. Even after a century, changes in the landscape may not be noticeable. Hillers and Beaman were photographers with the Powell expeditions of 1869 and 1871. They documented much of the scenery along the courses of the Green and Colorado rivers. A century later, Stephens returned to many of the early photographic sites to take modern photos and to look for changes in the land and vegetation. After 100 years, boulders and stones had not moved. Even piñon and juniper trees had barely grown. Most changes in this arid land are too slow for us to see.

SUGGESTED READING

Baars, Donald L. *Canyonlands Country.* Lawrence, Kansas: Cañon Publishers, Ltd., 1989.

Baars, Donald L. *The Colorado Plateau: A Geologic History.* Albuquerque, New Mexico: University of New Mexico Press, 1983.

Barnes, F. A. *Canyon Country Geology.* Salt Lake City, Utah: Wasatch Publishers, Inc., 1978.

Lohman, S. W. *The Geologic Story of Canyonlands National Park.* Washington, D.C.: U.S. Government Printing Office, Geological Survey Bulletin 1327, 1974.

Stokes, William L. *Geology of Utah.* Salt Lake City: Utah Museum of Natural History, 1987.

RUSS FINLEY

Along the northeastern edge of the *White Rim, a section of Organ Rock Shale has collapsed leaving a narrow ribbon of White Rim Sandstone—Musselman Arch. Slopes of the Moenkopi and Chinle formations lead to the Wingate and Kayenta cliffs on the horizon.*

Mesa Arch, at the end of a short trail *on the Island in the Sky, is in Navajo sandstone. Beneath Mesa Arch, in the center is Washer Woman Arch.*

DAVID MUENCH

Canyonland's Arches

the **Wooden** *shoe* is a **fanciful** **product** of *erosion*

BILL RATCLIFFE

In a side canyon five miles up Horse *Canyon, Castle Arch arcs across the skyline and resembles a medieval flying buttress.*

BECKY & GARY VESTAL

Wooden Shoe *Arch—in the Needles District —is a fanciful product of erosion of Cedar Mesa Sandstone.*

As you walk through any of the plant communities in the park, look closely at untrammeled ground. You are likely to see a blackish-grey microbiotic crust covering the soil.

The Living Canyons

TOM TILL

Water carves the rocky *landscape and determines the pattern of life in canyon country. Here, in the moist oasis of Holeman Spring, maidenhair fern thrives in a sanctuary from aridity.*

Canyonlands is not only a land of rock. In concert with climate, the land supports a mantle of life. Patterns of temperature, soil, and especially moisture allow particular plants to survive. Just as water is the key agent of erosion in Canyonlands, water determines the distribution of life in the park. The availability of moisture is the most important factor that controls where plants survive. In turn, the distribution of plants directly affects the pattern of animal abundance.

Away from the rivers and the few oases formed by seeps and springs, water is in short supply. Less than nine inches of precipitation falls on Canyonlands each year; one-third of it as brief downpours in late summer. Because there is so much bare rock and little soil, much of the rain runs off; little soaks into the ground. High summer temperatures and low humidity increase evaporation, while frigid winter temperatures freeze any free water. Both extremes mean that less water, than supplied by meager precipitation, is actually available to support life.

Communities of Plants

The distribution of plant communities—grassland, desertscrub, piñon-juniper woodland, riverbank, and hanging gardens—depends on patterns of moisture availability established by landforms.

This black, corrugated cryptobiotic crust once *covered great expanses of soil in southern Utah. The thin, fragile layer of lichens, mosses, and cyanobacteria stabilizes the loose substrate and adds nutrients to the underlying soil. Many seeds, like those of Indian ricegrass shown here,can germinate in the protected crevices in the crust.*

GARY LADD

***Indian ricegrass** is common in the West and is an important food for wildlife. Small mammals and birds gather its seeds and larger mammals, such as mule deer and bighorn sheep, eat the seeds along with stems and leaves. Prehistoric people also gathered its large seeds.*

MIKE HILL

KEITH GUNNER

***Ephedra, also known as Mormon tea,** has jointed stems, scale-like leaves, and produces male and female cones. It can be brewed into a stimulating tea.*

Indian ricegrass, needle-and-thread, galleta, and grama are the most common grasses. Their dense, shallow root systems quickly intercept moisture before it percolates deep into the sand. Shrubs and trees cannot absorb rainfall as it percolates rapidly and therefore cannot survive in deep sandy soil.

For nearly a century, domestic cattle grazed the grassy flats in Canyonlands. Salt Creek, Devil's Lane, and Squaw Flat in the Needles District were the most heavily used. Cattle ate the most palatable plants first and, in turn, allowed other initially less abundant species to thrive. Over the years, the species composition of the grasslands changed. This is especially noticeable in Devil's Lane and Squaw Flat where exotic species, such as Russian thistle, alias tumbleweed, and cheatgrass, replaced the more appetizing plants like Indian ricegrass. Grazing ceased in the park in 1983, and given time, the grass-lands may eventually return to their original character.

Benches and gentle slopes with shallow, rocky soil usually support stands of blackbrush, a short, spiny shrub that looks barely alive. The few inches of regolith (broken, disintegrated rock) perched atop bedrock concentrates water in a thin layer, in which blackbrush roots can slowly absorb enough moisture to sustain the plant. Together with shad-scale (a shrub with pale grey-green leaves) and Mormon tea, blackbrush covers large areas throughout the park. While blackbrush is not a favorite food of cattle, it was browsed extensively by domestic sheep. Native bighorn sheep also feed on the shrub and their populations may have suffered in the past from competition with domestic sheep for food.

Blackbrush is a tenacious, slow-growing plant. It occasionally flowers and produces seeds, and then only after a wet spring. Even when it blossoms, the shrub is drab. Tiny, dull-yellowish blooms give rise to tiny seeds; however, even if the seeds germinate rodents generally eat the seedlings. Once blackbrush gets a foothold it can withstand the climatic extremes of Canyonlands. If the plants are killed by over-graz-

***Preceding pages: Monument Basin** and the White rim from the air. Photo by F.A. Barnes.*

GARY LADD

Narrow-leaved *yucca inhabits sandy soil while, in the background, blackbrush predominates on thin, gravelly substrate. The yucca's cream-colored flowers and stalks are edible. Strips of the fibrous leaves were used by prehistoric Native Americans for tying bundles and plaiting sandals and mats.*

Will this lone pinon *receive any water from the nearby pool? Probably not. The roots of this natural bonsai have grown deep into a crevice in the sandstone and are watered when rain and snowmelt are funneled into the crack. E.O. Beaman photographed a similar tree in the Maze area in 1871. After 100 years that tree has barely grown.*

DAVID MUENCH

Rabbitbrush begins to bloom *in late summer, and by fall the shrubs produce a lavish show of yellow blossoms. Deer sometimes browse its leaves and stems. One species—rubber rabbitbrush—was once studied as a source of latex.*

ing, fire, or human activity, however, they re-vegetate the land very slowly. During the uranium mining "boom" of the 1950s, prospectors carved roads and airstrips across stands of blackbrush. Most of these scars are still visible and are devoid of blackbrush. This is a good example of how sensitive and fragile this apparently rugged landscape really is.

Piñons and junipers need more water than grasses or blackbrush, but paradoxically these trees grow in habitats with little or no soil. How can trees survive on exposed bedrock? Cracks and fissures breach the solid surface of slickrock. As rain falls or snow melts, the moisture runs into the crevices like water running into a funnel. These clefts collect water and are oases for plants with roots that can twist and extend deeply into the cracks. Piñons, junipers, and a variety of shrubs—cliffrose, mountain mahogany, barberry, and snowberry—flourish in the fractured rock. Sometimes the cracks are covered by a thin layer of soil. Look closely, though, and you can see that the trees and shrubs occur in lines or rows defined by the pattern of fissures. While piñons and junipers occur together in most of the park, at drier, lower elevations junipers are often alone.

The sagebrush community inhabits the deep sandy soil on benches adjacent to canyons such as Salt Creek, Lavender, and Davis. Groundwater is often close to the surface in these thin strips of land,

How can **trees** survive with **little** or *no soil* — exposed on bedrock

***N**arrow bands of water-loving plants line the* *riverbanks of the Green and Colorado. Even though river levels drop and the mud dries and cracks, plants still bathe their roots in shallow ground water.*

and the roots of big sage, rabbitbrush, and greasewood have a nearly year-round supply of water. The pungent smell of sage after a rain is unique and many visitors to the arid West carry home an enduring memory of the scent.

On the rubble slopes and benches that fringe the Green and Colorado rivers and other major canyons, the shrubby shadscale and saltbushes dominate the vegetation. These benches, perched just above the abundant water flowing in the Green and Colorado, are isolated from groundwater and are, surprisingly, among the driest habitats in the park. Their low elevation and thus high temperatures dry the ground quickly. Clay and shale compose much of the soil and although deep layers of clay retain water, the clay binds the water so strongly that roots of most plants can't absorb it.

The abundant and continuous supply of water provided by the Green and Colorado rivers supports a slender swath of moisture-loving plants along the riverbanks. Cottonwood trees, willows, and tamarisks require ample water and only survive where their roots bathe in plentiful groundwater. These strips of riparian vegetation, bordering the river courses, are narrow. The rivers lie in deep narrow canyons, and plants—only a stone's throw from the bank—may not have root systems deep enough to tap the underground water. Furthermore, riverside vegetation is not continuous along the river. In many places bedrock or rubble slopes reach all the way to the river's edge and plants cannot gain a foothold.

The tamarisk is a feathery-branched tree with plumes of pink and purple flowers. Although it is showy and attractive, this is an exotic plant in the United States and does not belong here! It was imported from western Asia to the Southwest in the mid-1800s as an ornamental shrub and for erosion control. Unfortunately, tamarisk was infinitely successful in its new home. Its high reproductive rate, combined with effective seed dispersal, resulted in its spread into most southwestern river drainages in only a few decades.

GARY LADD

***C**ottonwood, willow, and tamarisk along the Green* *River attest to the abundant supply of water close to the river. In contrast, none of the ground water is available to plants on the rubble slopes below the White Rim Sandstone.*

At Canyonlands, no tamarisk inhabited the canyons in 1920. Ten years later, this prolific shrub had populated the river canyons all the way upstream to the towns of Moab and Green River. Tamarisk has invaded and taken over much of the riverside habitat previously occupied by willows and cottonwoods. These native trees often cannot compete successfully with the aggressive tamarisk and are eliminated.

Many animals also lose their homes to tamarisk. The natural riverside communities evolved over millions of years and interactions among the plants and animals resulted in complex ecological relationships. Diversity of animals—especially birds, mammals, and insects—was high. When tamarisk moved in and replaced native vegetation, the balance of these ecological associations was disrupted by far less diverse and biologically interesting habitat than the missing native community.

In alcoves and beneath overhanging ledges where water seeps from between layers of rock dwell the rarest of plant associations in the park—hanging gardens. Protected from the sun and luxuriating in cool spring water are species unexpected in the desert: columbine, monkeyflower, death camas, primrose, and maidenhair fern. These plants depend on the moisture and shade of these scattered oases and are found nowhere else in the park. Because these fragile sites are often difficult to reach, they are mostly undisturbed.

As you walk through any of the plant communities in the park, look closely at untrammeled ground. You are likely to see a blackish-grey microbiotic crust covering the soil. This lumpy layer formerly was called cryptogamic soil and consists mostly of lichens, cyanobacteria (blue-green algae), and scattered mosses. Soil adheres to the cryptobiotic mantle and the loose surface of the ground is bound together. This living mat protects the underlying soil from erosion by wind and water, retards evaporation, and adds some nutrients to the infertile sand. Seeds often find sheltered cracks in the crust and plants may take root in the tempered microclimate of the layer.

As tough as the cryptobiotic crust looks, it is actually fragile and disintegrates readily underfoot. Mountain bikes and motorized vehicles traveling off roads, and hikers wandering from trails destroy this delicate mantle. Once the thin shield is breached, wind and rain tear into the unstable substrate and carry away the soil. After even a single passage, vehicle tracks and hiking routes across the soil are visible for decades. Thus the importance of the rangers' plea, "Please stay on trails and roads."

MIKE HILL

Hanging gardens occur where *seeps and springs emerge into the shelter of a rock overhang. Plants unexpected in arid canyon country—maidenhair fern, orchid, and monkeyflower—luxuriate in the cool, moist microhabitat of the hanging garden.*

DIANE ALLEN

Lavish flowering stalk of yucca.

DAVID MUENCH

A twisted and weathered juniper overlooks the Green River.

The Flora of Canyonlands

From moist riverbanks to arid slickrock seemingly inhospitable to life, the plants of Canyonlands encounter extreme habitats. Along the rivers and where scattered springs wet the ground, plants avoid the rigors of the desiccating environment. Only a few yards away, however, the ground is dry and only those plants adapted to the shortage of water can survive.

LARRY BURTON

The delicate purple blossoms *of the fishhook cactus contrast with its formidable hooked spines.*

THEA NORDLING

Butterfly milkweed is one of at *least five species of milkweed in Canyonlands. All are perennials with milky juice.*

GARY LADD

Decomposers, like this *mushroom, are important members of biological communities.*

Animals of Deserts and Mountains

While plants are anchored to the land and cannot escape the scorching sun or frigid wind, animals, on the other hand, are mobile. They can move within and between habitats if environmental conditions are not suitable. The simple act of going from a sunny rock to the shade of a shrub might be the difference between heat stroke and survival for a lizard. This freedom to move between "microclimates" is the key to survival for many Canyonlands animals. Most mammals are active at night and elude extremes of temperature, wind, and dry air by staying in burrows or rock shelters during the day. Many kinds of birds and bats inhabit Canyonlands only part of the year and migrate elsewhere when seasons change.

Animal life in Canyonlands is an interesting blend of mountain and desert species. Remember that the Colorado Plateau is not only arid slick- rock country. Forested highlands fringe the plateau and several mountain ranges punctuate the mile-high region. Indeed, the southern edge of Canyonlands is the 7,000-foot-high Salt Creek Mesa which leads to the Abajo Mountains. The fauna of the park has a component of species that evolved in deserts and a complement that originated in montane and highland environments. For example, in the sandy grassland of Squaw Flat, desert-adapted kangaroo rats hop in search of seeds, while black bears might wander into upper Salt Creek. Animal life is diverse—from tiny biting midges that harass hikers, to elusive desert bighorn sheep.

Some animals are tied closely to particular types of vegetation. The sage sparrow and piñon mouse are named so well that you instantly know where they typically occur. The sparrow forages and nests in stands of sagebrush. The piñon mouse lives in piñon and juniper trees and rarely ventures into another habitat. The two woodland trees provide seeds, leaves, and insects for food, nesting sites in the branches and cavities in the trunks, refuge from predators, and shelter from the elements. All the mouse's needs are met.

Other animals are not limited to a specific type of vegetation, but are tied to particular physical characteristics of the habitat. Canyon mice inhabit rocky slopes, cliffs, and barren slickrock flats. Wood rats search for cracks and ledges in cliffs where they build jumbled houses of sticks, rocks, and whatever else they can carry. Canyon wrens trill musically from perches on the sheer canyon walls. Bighorn sheep prefer rugged, isolated cliffs and talus slopes.

Near one of the campgrounds or along a short trail can be the best place to observe wildlife. Perch quietly on a rock or in the shade of a piñon. Watch for the quick, nervous scampering of the white-tailed antelope squirrel. This little ground squirrel darts out of its cool burrow or from the shade of a shrub to gather seeds. Its broad tail flips over the squirrel's back like an umbrella and provides a bit of protection from the sun. After filling its cheek pouches with seeds of Indian ricegrass or saltbush, the rodent dashes back to its nest to unload the cargo. This squirrel can be active in the heat of the day because

PHOTOS BY DAVID W. JOHNSON

Ord's kangaroo rat is supremely adapted to *arid, sandy habitats. It hops on strong hind legs and gathers seed with its short, dextrous forepaws.*

The piñon mouse is easily identified by its long *tail and inch-long ears. Because this mouse usually nests and feeds in junipers, perhaps it should be called the juniper mouse!*

Cliff swallows build nests of mud with *bits of plant material under the protection of overhanging ledges. Often they line the nests with feathers. Swallows are strong fliers and capture winged insects in their wide mouths.*

MIKE HILL

it alternates between sun and shade. If the animal gets too hot, it simply runs back to shelter to cool off before its next foraging trip.

The Colorado chipmunk and the larger grey-spotted rock squirrel are also active during the day, but tend to avoid the heat of midday. Both forage mostly in the cooler hours of morning and evening, while taking siestas during the most intense sunshine.

Although the turkey vulture, golden eagle, and raven wait for warm updrafts to support their daytime soaring searches for food, morning and "evening are the best times to see most birds. Listen for the chattering of titmice and gnatcatchers in the piñons and junipers, and the whistled song, "drink-your-tea," of the rufous-sided towhees in the Gambel oaks.

Nearly anytime during a summer day, white-throated swifts and violet-green swallows shoot like arrows past the cliffs. From Confluence Overlook and Grand View Point, these aerial acrobats dart and flutter on updrafts. At speeds over 100 miles per hour, swifts and swallows catch and eat insects in midair. These birds rarely seem to land, but when they do, they zoom at daredevil speed toward cliffs and pull up just in time to land on tiny nests tucked in crevices.

While many *homeotherms* ("warm-blooded" animals) try to avoid overheating by being active at night, *poikilotherms* ("cold-blooded" animals), like reptiles, need to bask in the sun and warm up to "operating" temperature before they can hunt. Many of the lizards in the park perch atop boulders and rocky outcrops. From these sunny vantage points, they proclaim their territories, attract mates, look for food, and absorb warmth from the sun. Of course, after they get warm, these reptiles need to avoid getting too hot and will wander back and forth from sun to shade to maintain appropriate body temperature.

MIKE HILL

Listen for the haunting howls and yelps of the *coyote anywhere in Canyonlands.*

Animal life is an ***interesting*** **blend** of ***mountain*** and ***desert*** *species*

DAVID W. JOHNSON

***S**crub jays are raucous camp robbers in canyon country. They supplement their natural diet of plants and animals with scraps of food left accidentally by park visitors. Piñon jays also are common in Canyonlands and are distinguished by their uniform dull blue plumage.*

All the lizards in the park are carnivores. The little side-blotched lizard and tree lizard eat mostly insects. The larger fence and whiptail lizards supplement their insect diet with the smaller lizards. The foot-long leopard and collared lizards specialize on their smaller relatives. From a viewpoint high on a boulder, these aggressive predators easily overpower smaller lizards and even mice.

If you visit Canyonlands or just about any other wild area on the Colorado Plateau during late May or early June you will probably encounter one member of the biota that you will wish you had not—biting midges. These tiny, gnat-sized flies emerge for a few weeks at the beginning of summer and feed on nectar and blood! Insect repellent does not deter their attacks and all you can do is wait until the cool of evening when they relent.

In contrast to the midge is the bighorn sheep. This graceful mammal is at home on apparently sheer cliffs and steep talus slopes. Both sexes have horns, but only males grow the massive, spiraling ones. Unlike the deciduous antlers of deer, horns are permanent and their ever-increasing curl reflects the animal's age. Bighorns prefer to eat grasses and other herbaceous plants, but will survive by browsing on blackbrush and other shrubs. Regardless of diet, they feed near the protection of cliffs and rubble slopes where they can flee from predators. Mountain lions, bobcats, and coyotes can kill bighorns, but diseases probably reduce their populations more than predators.

In the past, bighorn sheep were abundant over much of the Colorado Plateau. But because of disease and competition for habitat with domestic livestock, their abundance declined dramatically. Bighorn populations in the Island in the Sky District of Canyonlands are healthy and some individuals have been removed from the park to reestablish populations in other suitable areas of southern Utah. Because of the bighorn's generally shy nature and remote habitat, backcountry travelers and river runners are usually the only visitors fortunate enough to see them.

Several rare and endangered species of animals find refuge and protection in Canyonlands. The Colorado River squawfish, humpbacked chub, and bonytail chub are once-abundant fish that are now threatened with extinction. Introduction of non-native fish and construction of dams and reservoirs on

THEA NORDLING

***T**his inch-long crustacean, called a tadpole shrimp, inhabits the ephemeral pools of water in slickrock called potholes. Spring and summer rains fill shallow impressions in the sandstone and as sunlight warms the water, eggs of invertebrate animals hatch and snails buried in the muddy substrate "awaken" from their dormant state. Potholes soon teem with life as many species race to reproduce before the pool dries.*

***Bighorn sheep are at home on rough,** rocky slopes and in backcountry canyons. They clamber with ease over apparently impassable terrain. Rams display massive curling horns while ewes grow shorter, slightly curved spikes. Bighorns breed in winter and ewes give birth in early summer. Although predators kill bighorns, diseases such as lungworm and sinusitis also affect populations.*

LARRY BURTON

the Green and Colorado rivers have changed aquatic habitats to the extent that these native species may not survive. The combination of changes in the physical environment of the rivers—silt load, temperature, bottom texture, rate of flow—and predation and competition by exotic species have altered the ecological rules-of-the-game, and the squawfish and chubs may be the losers. On a happier note, two rare species may be doing well in the park. The peregrine falcon appears to be taking advantage of the remote cliffs and reestablishing itself in part of its former range in Utah. The elusive spotted owl has never been abundant in the Southwest, but may be holding its own in the moist canyons of the park.

As you look at the landscape, in the heat of summer or in the biting wind of winter, consider the diversity of life that inhabits Canyonlands. Even the most rigorous or seemingly inaccessible places are inhabited. The summits of isolated buttes and mesas support populations of mice, birds, reptiles, insects, and plants. Even solid, exposed sandstone is home to lichens. Canyonlands is indeed a showcase for the vitality and tenacity of life.

SUGGESTED READING

ABBEY, EDWARD. *Desert Solitaire.* New York, New York: Simon and Schuster, 1968.

ARMSTRONG, DAVID M. *Mammals of the Canyon Country.* Moab, Utah: Canyonlands Natural History Association, 1982.

FAGAN, DAMIAN. *Canyon Country Wildflowers.* Helena Montana: Falcon Press, 1998.

WILLIAMS, DAVID. *A Naturalist's Guide to Canyon Country.* Helena, Montana: Falcon Press, 2000.

Among these [towers] by far the most remarkable was the forest of Gothic spires... Nothing in nature or in art offers a parallel to these singular objects...

DR. JOHN S. NEWBERRY

Before there was a Park

Human history in the canyon country is ancient. While we might feel like the first to wander remote canyons, people have been visiting Canyonlands for millennia. As far back as 10,000 years ago, bands of prehistoric hunters searched the Colorado Plateau for giant bison, elk, and other large prey. These Paleo-Indians were nomadic and followed herds of large mammals. There is some evidence of Paleo-Indian use of Canyonlands, and their beautifully crafted stone spear points have been found in the park.

DAVID W. JOHNSON

***The kiva was a vital structure in the Anasazi** culture where many religious ceremonies were conducted. This view into the chamber is through the entrance in the roof.*

CANYON ANCIENTS

Paleo-Indian tradition thrived from about 11,000 B.C. to about 5000 or 6000 B.C. By 4000 B.C., the post-glacial climate and wildlife of North America had developed, and the culture of the descendants of Paleo-Indians had changed to the extent that archaeologists refer to the modified culture as Archaic.

Archaic people still hunted, but some of the big game of their ancestors was extinct and the Archaic hunters relied on a variety of animals. They supplemented their arsenal of spears with other weapons like atlatls (throwing-sticks), snares, and nets. Plants were also an important part of their diet. Like their ancestors, the Archaic people were nomadic. As seasons changed and the game animals migrated or became scarce and as different plant foods were available, the people moved on.

By about 1000 B.C., however, the Archaic people began to grow maize (corn). Cultures in Mexico domesticated maize several thousand years earlier, and slowly this knowledge spread north. Several hundred years after assimilating corn into their diet, residents of the Southwest added domestic beans and squash. At first these new sources of food were only supplements. Corn could be planted in spring near a good winter camp and left with little care

JOSEF MUNECH

***If you drive to the Needles District along** Indian Creek, you will pass Newspaper Rock. For millennia, people noted their passage on the protected face of this cliff. Ancient designs, images of animals, and hand- and footprints mingle with historic petroglyphs of hunters on horseback.*

BECKY & GARY VESTAL

***The Great Gallery is a major panel of** pictographs in horseshoe Canyon, a detached unit of Canyonlands to the west of the Green River. Figures that are larger than life loom above the canyon bottom. Most of the figures lack extremities and most were painted in red, some with white. Some have imposing skull-like visages that add to their mysterious appearance. These pictographs predate Anasazi and Fremont occupation of the region and are at least several thousand years old.*

while the people followed their traditional foods. Any corn that had survived, upon their return to camp in the fall, was a bonus and would provide a little extra food for the winter.

Over time the people gave more attention to their crops and relied on them more. Since agriculture and nomadism are not really compatible, the Archaic people became dependent on their crops and gradually became sedentary. They built permanent villages and eventually supplemented their diet of domesticated crops with wild vegetation and game. This transition in lifestyle from hunting and gathering to agriculture was significant. Archaeologists have designated this new cultural tradition in the Four Corners area as Anasazi.

Anasazi

Throughout the national parks of the Southwest, evidence of Anasazi culture is astounding: hidden cliff dwellings of Keet Seel, towers of stone at Hovenweep, 600-room pueblos of Chaco Canyon, wondrous architecture at Mesa Verde. For more than 1,200 years the Anasazi flourished in the canyons and on the mesas. Their culture was sophisticated and not provincial. They traded pottery and turquoise for feathers and shells with peoples of Mexico and western America. Their population grew and by A.D. 1100, people of Mesa Verde had established villages as far north as Canyonlands.

At the same time the Anasazi culture was developing in the Four Corners region, the Archaic tradition was giving way to the Fremont culture to the north and west in Utah and Nevada. While the origins of the Fremont tradition are poorly understood, archaeologists have suggested several hypotheses to account for the Fremont's prehistoric roots. Some speculate that the Fremont people had roots in the Great Basin, others say that they were early offshoots of the Anasazi. Whatever their lineage, the Fremont and Anasazi peoples were contemporaneous in the Canyonlands region.

Northern expansion from the Mesa Verde area in southwestern Colorado led Anasazi people, in about A.D. 1100, to what became their northernmost outposts in the Needles District of Canyonlands. Just to the north and west of the Needles, across the Colorado River, lived the Fremont. It appears that the Colorado River formed a loose boundary between the groups.

The Salt Creek drainage in the Needles District contains many Anasazi habitation and rock art sites, and is listed on the National Register of Historic Places as the Salt Creek Archaeological District.

The legacy of the Fremont people is less well understood. However, just as the Anasazi decorated cliffs and boulders with paintings and etchings (pictographs and petroglyphs), the Fremont also developed a distinctive style of art. In all three districts of the park, there are panels of rock art that depict hunting scenes, crop harvesting, stylized figures, and abstract designs.

There are many Fremont-style rock art panels in the typical Anasazi territory of the Needles District. Often one can find a ruined stone dwelling originally built and occupied by the Anasazi, and close by see a Fremont panel of shield figures and fat bighorn sheep. Archaeologists suggest that the paintings were drawn by Fremont or that the Anasazi adopted the style of art from their neighbors. We will probably never know who decorated the canyon walls or why they did so, but we certainly can appreciate the skill and artistry of some of these compositions.

In the upper reaches of Salt Creek, there remain many ruins of small Anasazi habitations. Because they needed access to water for farming, families built homes and storage structures (granaries) in the canyons with at least seasonal creeks. There is no record of large pueblos in Canyonlands. Because there was only enough arable land to support small groups of people, most of the Anasazi dwellings we find could have housed one to three families.

The largest cliff dwelling in the park, in the upper reaches of Salt Creek, is informally called "Big Ruin." It contains two dozen structures, including dwellings and granaries. A few other modest-sized sites occur in the park, but most are small, probably representing seasonal use of the area.

In addition to the crops, wild plants, and animals that sustained the Anasazi, the people raised turkeys. Feathers were probably used for decorations, to insulate clothes, and perhaps in ceremonies. Even turkey guano could have fertilized the fields.

These prehistoric peoples had a strong impact on the land. They converted canyon bottoms into fields, hunted large and small animals, gathered plant foods, and cut piñon and juniper trees for building materials and firewood. Although the population was probably never large, after a century or two of occupation the people may have exceeded the ability of their habitat to sustain them. Firewood and game may have become scarce. Decades of farming may have depleted some nutrients from the soil.

As the Anasazi's impact increased, the climate also turned against them. A 20-year drought in the 13th century affected nearly all the people of the Southwest. In Canyonlands, where subsistence must have been marginal at best, the lack of resources, combined with too little precipitation, may have forced the Anasazi to leave and migrate to the south and east in search of better living conditions. By the end of the 13th century, Canyonlands was abandoned. The Fremont and Anasazi alike had left their homes, never to return.

The largest cliff dwelling in the park is in upper Salt Creek. Perched on a ledge, more than a dozen stone structures provided shelter and storage rooms for the Anasazi. Many smaller sites occur in the labyrinth of canyons and side canyons in the Needles.

K.C. DENDOOVEN

F. A. BARNES

One of the Sixshooter Peaks rises along the *road into the Needles District. With its twin, it was named for its resemblance to "six-guns." The spire is an erosional remnant of a once-larger mesa. All that remains is a sliver of Wingate Sandstone.*

During the subsequent 200 years, this land was visited only seldom by other groups of Native Americans. While Navajos and Utes (newcomers to the region) probably ventured into the rugged canyon country in search of game, they left little evidence of permanent habitation. Like the ancient cultures, they too left their calling cards on the rocks.

Explorers

The first European penetrations of the Southwest were by the Spanish soldier and explorer, Francisco Vasquez de Coronado. In 1540-1541, Coronado led his troops and auxiliaries into what is now New Mexico, but came nowhere close to Utah. It would be another 200 years before Europeans pushed their frontier into southeastern Utah and Canyonlands.

In 1765, Juan Maria Antonio Rivera explored from northern New Mexico into southern Colorado and southeastern Utah. He and his men were the first non-Native Americans to see the spectacular ruins of Mesa Verde and Hovenweep. They found a route across the La Sal Mountains and camped along the Colorado River near the present-day town of Moab. But they did not wander down the Colorado into the rugged canyonlands.

A decade later, two Franciscan priests left Santa Fe, New Mexico, in search of a trade route to the coast of California. Fray Silvestre Vélez de Escalante and Fray Francisco Atanasio Dominguez set out with a few merchants and workers in late summer 1776. They crossed the Colorado River near the present location of Grand Junction, Colorado, and reached north-central Utah as winter was coming. Rather than face a bitter winter in unknown country, the expedition headed back towards Santa Fe and passed to the west of Canyonlands. All of these early expeditions avoided the heart of the canyon country.

In 1836, Denis Julien first passed through the rugged canyon country that others had skirted. Julien carved his name, initials, and the date of passage into boulders and cliffs along the Green and Colorado rivers from Desolation Canyon to Glen Canyon. Why was he roaming through this unexplored territory? The scant records of his activity indicate that he was a fur trapper—perhaps he was searching for unexploited populations of beavers. Finally, in 1859, an organized effort was made to examine the heart of the unknown canyon country. Captain John N. Macomb commanded an expedition to locate the confluence of the Green and Grand (as the Colorado was then called), to determine the course of the San Juan River, and to find the most direct route from the Rio Grande of New Mexico to the settlements in southern Utah. In August 1859, Macomb reached a high point above the Colorado River in the present-day Needles District and scanned the panorama that extended as far as the eye could see. He was a practical person and was hardly captivated by the spectacular

DIANE ALLEN

Denis Julien was a trapper who traveled along *the Green and Colorado rivers in the 1830s. He left several records of his passage on cliffs.*

***A maze of canyons and cliffs discouraged extensive** exploration of the region. When ranching developed in the area, cowboys had no choice but to wander the canyons in search of water, grazing land, and stray livestock. Foot and horseback were (and probably still are) the best modes of transportation in canyon country.*

F. A. BARNES

scenery. In his final report, he remarked, "I cannot conceive of a more worthless and impracticable region than the one we now found ourselves in." Not the sort of endorsement expected for a region destined to become a premier national park!

Not all the members of Macomb's group were as dismayed by the forbidding nature of the land. Dr. John S. Newberry accompanied the expedition and, unlike Macomb, was enthralled by the landscape. Looking into the Needles District, Newberry wrote, "Among these [towers] by far the most remarkable was the forest of Gothic spires, first and imperfectly seen as we issued from the mouth of the Cañon Colorado. Nothing I can say will give an adequate idea of the singular and surprising appearance which they presented from this new and advantageous view.... Nothing in nature or in art offers a parallel to these singular objects...."

The most famous and certainly the most scientific of the early expeditions were the two journeys of John Wesley Powell and his comrades. In 1869 and again in 1871, Powell led two small groups of men down the Green River to its confluence with the Colorado and downstream through Cataract, Glen, and Grand canyons. Powell's expeditions provided the first detailed geologic and topographic information on these canyons and filled in some large blank expanses on the map.

E. O. Beaman was a photographer on Powell's second venture and, nearly a century later, H. G. Stephens found many of Beaman's camera locations and duplicated the original photographs. Side-by-side comparisons of the pairs of images provide vivid evidence of the magnitude of geologic time. The photos show that, even in 90 years, stones and boulders have not moved, piñons have barely grown; the land has not aged.

Powell's success showed that the great canyons could be traversed—that they would possibly provide routes to mineral-bearing deposits and even paths for railroads. Perhaps the most ambitious plan was Frank M. Brown's proposal to build a transcontinental rail line, the Denver, Colorado Canyon, and Pacific Railroad, using the canyons of the Colorado as the route. Early surveys reconnoitered the easy, flatwater stretches and encouraged the group to continue. But the surveyors made their way with great difficulty through Cataract Canyon and then met with tragedy in the upper reaches of Grand Canyon. Brown and two others drowned and the expedition was aborted.

During the late 1800s and into the early 1900s, the river was mainly a means of access for a few hardy prospectors. Best, Hislop, Galloway, and Loper are just a few of the miners who eked a living from the gold in the sands of the Colorado River.

Cowboys and Prospectors

While river runners tried to exploit the resources of the Green and Colorado rivers, the land above the rivers was also being used. The range cattle industry invaded southeastern Utah slowly at first in the 1870s. But by the early 1880s, over 10,000 cows roamed the open range around Monticello, Blanding, and Moab. In 1885, Turner and Cooper established what became known as the Dugout Ranch along Indian Creek, and were probably the earliest to graze

***The Colorado River** changed from a barrier to travel to a pathway through the canyon country in the late 1800s when prospectors found gold in the sands of the river. Here, below Dead Horse Point State Park, the Colorado is smooth and tranquil. However, 41 miles downstream the river enters tumultuous Cataract Canyon. The rapids in 10-mile-long Cataract Canyon are some of the wildest in North America.*

cattle in Canyonlands. Others—Goudelock, Titus, Kirk—joined them in the area, and by the end of the century much of the range was overgrazed.

Two brothers, Bill and Andrew Somerville, joined forces with Al and Jim Scorup and took over the Indian Creek cattle operation in 1919. The S & S Cattle Company was soon the biggest ranch in Utah, with over 10,000 cattle using 1.8 million acres of grazing land. The S & S faltered in 1967 and was bought by Robert Redd, who revitalized the ranch. As you drive down Indian Creek to the Needles, you pass near the headquarters of Dugout Ranch.

The history of cattle and sheep on the Island in the Sky and in the Maze follows a similar pattern, only the names are different. Taylor, Murphy, Ekker, Holeman, Tibbets, and Holyoak are names of early cattle ranchers that live on in Canyonlands. The Ekker family still ranches the open grasslands west of the Green and Colorado. The others are remembered in the area's place- names, and have proud descendants living in the canyon country.

Some of the people of this cowboy period were famous...or perhaps infamous. Butch Cassidy and the Wild Bunch used Robber's Roost Canyon, to the west of the park, as a hideout from 1884 to about 1900. Other characters are hardly known. In a remote area of the park north of Lockhart Canyon, an unknown cowboy is remembered only by his name and date of death—"Butch Midrid, 18??-1911"—scratched on a lonely upturned rock.

World War II had a surprisingly strong impact on the canyon country. The need for new sources of oil and gas, the development of nuclear weapons, and a budding nuclear power industry stimulated the last great opening of the Canyonlands region. In the 1950s and '60s, prospectors in jeeps and on foot searched the Chinle Formation for uranium. Roads were bulldozed across the land. Several deep exploratory wells were drilled in what is now the park in search of oil and gas. Hundreds of small pits were dug into slopes of shale and mudstone by prospectors eager to find the "motherlode" of uranium. The ore was common in the park, but not in rich deposits. The natural riches of the Canyonlands would ultimately not be found in the exploitation of its minerals or wildlife, but in its exquisite beauty.

The muddy waters of the Colorado River *join those of the Green River near the center of Canyonlands. The waters don't mix quickly, but mingle gradually downstream. John Wesley Powell and his comrades camped for several days at the confluence in 1869.*

SUGGESTED READING

BARNES, F. A. *Canyonlands National Park: Early History and First Descriptions*. Moab, Utah: Canyon Country Publications, 1988.

CRAMPTON, C. GREGORY. *Standing Up Country: The Canyonlands of Utah and Arizona*. New York, New York: Alfred A. Knopf, 1964.

JONES, DEWITT, AND LINDA S. CORDELL. *Anasazi World*. Portland, Oregon: Graphic Arts Publishing Company, 1985.

LAVENDER, DAVID. *Colorado River Country*. New York, New York: E. P. Dutton, Inc., 1982.

MARTINEAU, LA VAN. *The Rocks Begin to Speak*. Las Vegas, Nevada: KC Publications, 1973.

NEGRI, RICHARD F. *Tales of Canyonlands Cowboys*. Logan, Utah: Utah State University Press, 1997.

SCHAAFSMA, POLLY. *Indian Rock Art of the Southwest*. Santa Fe, New Mexico: School of American Research, 1980.

Turk's Head is an isolated cap of White Rim *Sandstone at a bend in the Green River.*

People at Canyonlands

Canyonlands has been visited by people for millennia—from Paleo-Indians pursuing prey to modern-day adventurers. As a national park, Canyonlands preserves a portion of the wild canyon country and provides a variety of paths to experience its natural legacy.

***The road over Elephant Hill began as a route to** drive livestock to grazing areas. During the uranium boom of the 1950s, the way was widened. The road is still a gateway to the Needles backcountry and tests the mettle of four-wheel-drive vehicles and drivers alike!*

KEN MABERY

***Traveling the rapids of Cataract canyon for fun** probably began with Nathaniel Galloway's trips at the turn of the century. Today, hundreds of river runners traverse the canyon every year.*

TOM COX

KEITH GUNNAR

***Mountain** bikers have discovered the roads of Canyonlands. Quiet and non-polluting, bikes are an ideal way to visit the park.*

TOM TILL

***Along the Cave Spring Environmental Trail is an** abandoned cowboy camp. Notice the metal cans on the table legs—attempts to keep mice off the table.*

Bates Wilson

Bates Wilson began his career in the National Park Service as acting superintendent at Organ Pipe Cactus National Monument. In 1949 he became superintendent at Arches National Monument (as it was called then). After moving to Arches, he explored the land destined to become Canyonlands National Park on foot, horseback, and by four-wheel-drive vehicle. He loved showing that wondrous land to friends and visitors. By far, he enjoyed cooking them a dutch-oven meal more than dealing with his bureaucratic duties. In 1964, he became the first superintendent of Canyonlands and remained at that post until he retired in 1972. After retirement, Wilson worked a ranch in Professor Valley (up the Colorado River from Moab, Utah) until his death in 1983.

NPS PHOTO BY WOODBRIDGE WILLIAMS

KEITH GUNNAR

***Every year more visitors** discover the wonders of Canyonlands. There is a diversity of marked trails in the park, from short walks to long, strenuous hikes.*

All About Canyonlands National Park

Canyonlands Natural History Association

Canyonlands Natural History Association founded in 1967, is a non-profit organization established to assist the scientific and educational efforts of the National Park Service, Bureau of Land Management and USDA Forest Service, agencies that together oversee more than 7.5 million acres of federal land in southeast Utah. Their goal is to enhance each visitor's appreciation of public lands by providing quality educational materials, both free and for sale.

Revenues are used to support the agencies' programs in various ways, including seminars, outdoor educational programs for area schools, equipment and supplies for ranger naturalists, exhibits, new facilities and funding for research.

Contact Us

Address
Canyonlands National Park
2282 S. West Resource Blvd.
Moab, UT 84532-3298

Telephone
Vistor Information
435-719-2313

Visitor Information Email
canyinfo@nps.gov

SCRUB JAY PHOTO BY DAVID W. JOHNSON

CANYONLANDS *Junior Ranger*

Are you between the ages of six and twelve? Do you want to learn more about Canyonlands National Park?

Become a Junior Ranger, and find out what different jobs the Park Rangers perform. Travel around the park and explore the different sites, and see what animals reside within the park boundaries. Learn about Geology, the Ancestral Puebloans, and all about the pioneers.

Pick up your Junior Ranger booklet from the visitor center, complete at least five of the activities in the booklet, and return to speak with a Park Ranger. The Ranger will sign your certificate and pin on your official Junior Ranger badge! **Wear it with pride!**

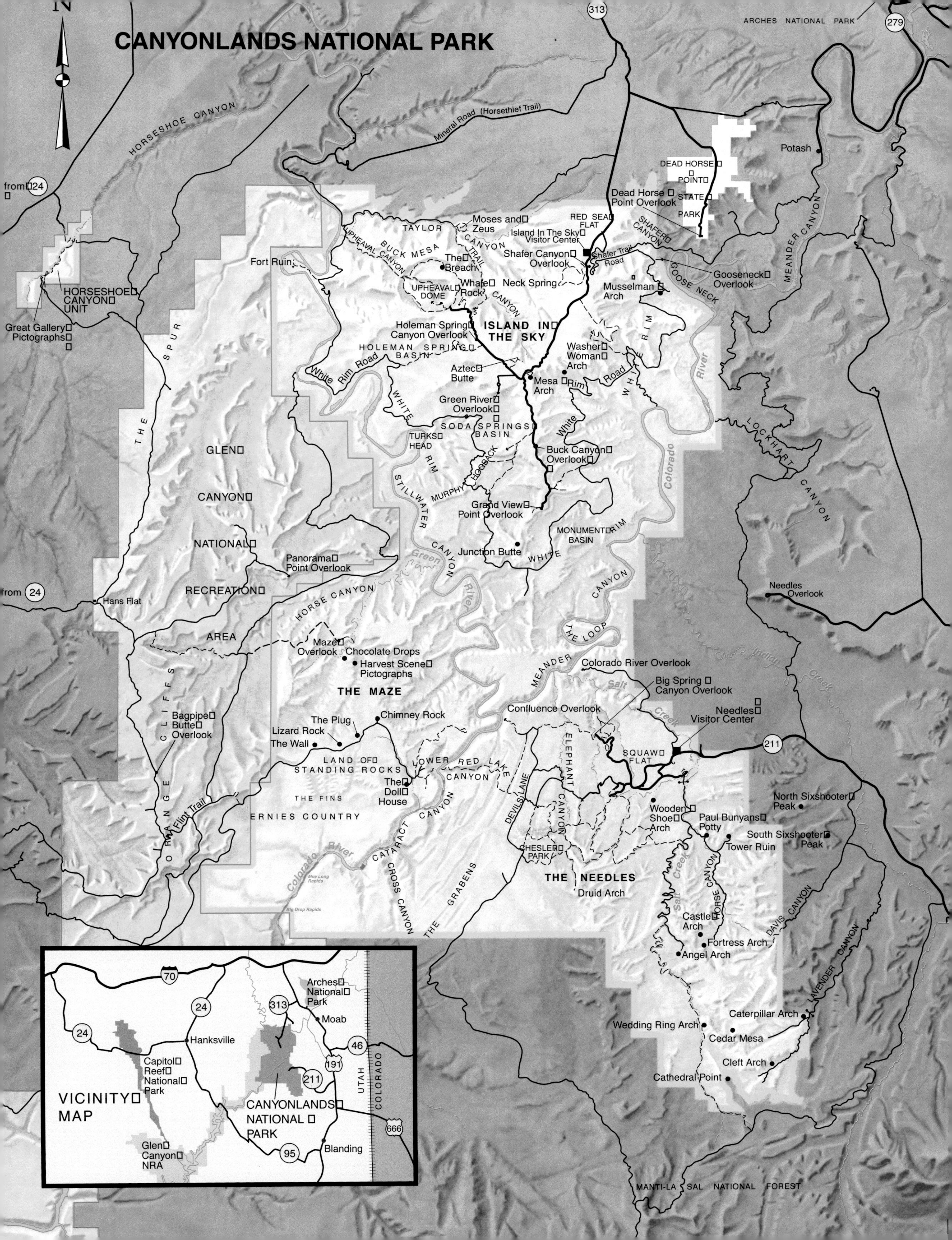
CANYONLANDS NATIONAL PARK
N
ARCHES NATIONAL PARK
313
279
HORSESHOE CANYON
Mineral Road (Horsethief Trail)
Potash
DEAD HORSE POINT STATE PARK
Dead Horse Point Overlook
from 24
RED SEAL FLAT
Moses and Zeus
TAYLOR
Island In The Sky Visitor Center
SHAFER CANYON
MEANDER CANYON
UPHEAVAL CANYON
BUCK MESA
CANYON TRAIL
Shafer Canyon Overlook
Shafer Trail Road
Fort Ruin
The Breach
Gooseneck Overlook
GOOSE NECK
Whale Rock
Neck Spring
Musselman Arch
UPHEAVAL DOME
HORSESHOE CANYON UNIT
Great Gallery Pictographs
Holeman Spring Canyon Overlook
ISLAND IN THE SKY
HOLEMAN SPRING BASIN
Washer Woman Arch
White Rim Road
Aztec Butte
Mesa Arch
THE SPUR
Green River Overlook
SODA SPRINGS BASIN
TURKS HEAD
LOCKHART CANYON
GLEN CANYON NATIONAL RECREATION AREA
Buck Canyon Overlook
Colorado River
MURPHY HOGBACK
STILLWATER CANYON
Grand View Point Overlook
MONUMENT BASIN
Panorama Point Overlook
Junction Butte
Green River
from 24
Hans Flat
Needles Overlook
HORSE CANYON
THE LOOP
Maze Overlook
Chocolate Drops
Harvest Scene Pictographs
Colorado River Overlook
MEANDER CANYON
Big Spring Canyon Overlook
Salt Creek
Indian Creek
THE MAZE
ORANGE CLIFFS
Bagpipe Butte Overlook
Confluence Overlook
Needles Visitor Center
Chimney Rock
The Plug
Lizard Rock
The Wall
211
LAND OF STANDING ROCKS
LOWER RED LAKE CANYON
ELEPHANT CANYON
SQUAW FLAT
The Doll House
THE FINS
North Sixshooter Peak
Flint Trail
ERNIES COUNTRY
DEVILS LANE
Wooden Shoe Arch
Paul Bunyans Potty
South Sixshooter Peak
Tower Ruin
CATARACT CANYON
CHESLER PARK
THE NEEDLES
Druid Arch
Mile Long Rapids
Big Drop Rapids
CROSS CANYON
THE GRABENS
HORSE CANYON
DAVIS CANYON
Castle Arch
Fortress Arch
Angel Arch
LAVENDER CANYON
Caterpillar Arch
Wedding Ring Arch
Cedar Mesa
Cleft Arch
Cathedral Point
MANTI-LA SAL NATIONAL FOREST
70
Arches National Park
313
Moab
24
24
Hanksville
46
Capitol Reef National Park
191
211
UTAH
COLORADO
VICINITY MAP
CANYONLANDS NATIONAL PARK
666
Glen Canyon NRA
95
Blanding

Canyonlands Today

With the enthusiasm, guidance, and foresight of people such as former Secretary of the Interior Stewart L. Udall, Bates Wilson, Slim Mabery, Kent Frost and many others, the heart of the canyon country was preserved as Canyonlands National Park. On September 12, 1964, President Lyndon B. Johnson signed legislation establishing the first national park since 1956. In 1971, President Richard M. Nixon signed into law an expansion of the park to its present size of 527 square miles.

Canyonlands National Park preserves only a fragment of the wild canyon country. Grazing, hunting, and exploration for minerals are no longer allowed, but like other national parks, Canyonlands is nevertheless threatened. Increasing visitation in the rugged, yet fragile land taxes the ability of the environment to recover from abusive off-road driving, mountain biking, camping, and cross-country hiking. From outside the domain of the park's authority, vista-impairing polluted air and degraded water flow in.

We need to be sensitive to the nature of canyon country and be aware that the national parks on the Colorado Plateau—like Arches, Bryce Canyon, Canyonlands, Capitol Reef, and Zion—are but remnants of a once-vast wilderness. Only our thoughtful actions in behalf of the national parks will preserve them unimpaired for posterity.

BECKY & GARY VESTAL

Cedar Mesa Sandstone glows in the light of evening.

Published by KC Publications, 3245 E. Patrick Ln., Suite A, Las Vegas, NV 89120.

***Inside back cover:** Washer Woman Arch on the Island in the Sky. Photo by David Muench.*

***Back cover:** The Colorado River continues its inexorable sculpting of the canyon country. Photo by Mike Hill.*

Created, Designed, and Published in the U.S.A.
Printed by Tien Wah Press (Pte.) Ltd, Singapore
Pre-Press by United Graphic Pte. Ltd